HYDROPONIC GARDENS

Your Way to Lush, Blooming Gardens!

A comprehensive manual outlining the sequential process of constructing a personalized hydroponic system.

Rita van Klaveren

Table of Contents

INTRODUCTION

Hydroponics is a soilless cultivation technique that involves supplying plants with a water-based solution containing vital nutrients, which are delivered straight to the roots.

Hydroponics expands the limits of your gardening expertise. Hydroponic systems offer limitless options for both experienced enthusiasts and interested beginners. Imagine the vivid hues of aromatic herbs, the juiciness of fully ripened tomatoes, and the crunchiness of verdant greens, all conveniently accessible and nurtured via your own cultivation.

However, hydroponics encompasses more than simply a

garden; it is a way of life. The concept revolves around re-establishing a connection with the natural world inside the midst of urban environments, converting balconies, rooftops, and internal areas into lush havens of peace and serenity. It pertains to relishing the flavor of locally cultivated crops and having precise knowledge of the origin and cultivation methods of your food.

Hydroponic solutions revolutionize ease and efficiency. Eliminate the need for laborious soil manipulation, combating pests, or concerns over capricious weather conditions. Hydroponics enables effortless and continuous gardening throughout the year, guided by accuracy and eco-friendliness.

Embrace the hydroponics revolution and harness the potential of hydroponics in your daily existence. Enhance your gardening expertise, nurture your own natural sanctuary, and embrace a more environmentally friendly and healthier future.

Are you prepared to commence this remarkable expedition into the world of hydroponics? Now, immerse yourself in this comprehensive guide!

CHAPTER 1

What is hydroponics?

Ancient civilizations like the Babylonians and Aztecs pioneered the origins of hydroponics. These civilizations used basic hydroponic techniques to grow plants in water-based settings. Yet, it was not until the 20th century that hydroponics became a feasible agricultural method, propelled by scientific and technological progress.

During the 1930s, scientists initiated research to investigate plant development in nutrient solutions, establishing the foundation for contemporary hydroponic systems. Researchers and farmers further improved hydroponic techniques by devising diverse methods and systems customized for specific crops and locations.

Hydroponics, the practice of cultivating plants without soil, is a cutting-edge technique in agriculture that provides inventive answers to worldwide food production concerns. Hydroponics, a fusion of traditional farming methods and advanced technology, revolutionizes crop cultivation by offering sustainable, efficient, and space-saving techniques to address the needs of our expanding global population.Hydroponics is based on the fundamental idea that plants obtain necessary nutrients not from soil directly but from water that contains dissolved minerals and other crucial ingredients. Hydroponic systems enhance plant growth by circumventing the conventional soil-based method and providing nutrients directly to the roots. This results in accelerated growth rates, increased yields, and improved management of crop quality.

Benefits of Hydroponic gardening

Hydroponic horticulture has a multitude of advantages compared to traditional soil-based agriculture, which has led to its growing popularity among farmers, academics, and urban gardeners. Several notable benefits include:

Enhanced Productivity: Hydroponic systems provide

meticulous regulation of environmental variables, including nutrient concentrations, pH levels, and temperature, leading to accelerated growth rates and elevated agricultural outputs in comparison to conventional techniques.

Water Efficiency: Hydroponic systems are more water-efficient compared to traditional farming methods due to their ability to recirculate nutrient solutions and minimize water consumption. This makes them especially suitable for dry regions and locations that are susceptible to drought.

Space Optimization: Hydroponic systems can be customized to suit various places, ranging from compact urban flats to expansive commercial greenhouses, optimizing crop output within confined regions and vertically arranged settings. This enables farmers to optimize crop output within confined regions and vertically arranged settings.

Minimized Environmental Impact: Hydroponic farming mitigates the need for chemical fertilizers and pesticides, hence diminishing soil erosion, water pollution, and other environmental effects linked to traditional agriculture.

Continuous Production: Hydroponics enables producers to sustain ideal growth conditions throughout the year, irrespective of seasonal variations and unfavorable weather circumstances. This guarantees a steady provision of fresh produce, regardless of the time of year.

Limitations and Factors to Take into Account

Although hydroponics presents numerous advantages, it also entails certain difficulties and factors to take into account. Several crucial things to take into account are:

Capital Outlay: The establishment of a hydroponic system necessitates an initial investment in physical infrastructure, specialized equipment, and advanced technology, which may provide a financial barrier for individuals engaged in small-scale cultivation or recreational gardening.

Technical Experience: Proficiency in hydroponic farming requires individuals to possess comprehensive knowledge of plant biology, nutrient management, and system dynamics, which they can acquire through specialized training and experience in hydroponic principles and methods.

Maintenance Requirements: Hydroponic systems necessitate consistent surveillance and upkeep to guarantee ideal nutrient concentrations, pH equilibrium, and system efficiency, which might demand substantial work and time investment.

Disease and Pest Management: Producers must still be cautious of airborne infections and pests that can endanger crops, despite the reduced vulnerability to diseases and pests transmitted through soil in hydroponic systems.

Fundamental principles of hydroponic gardening

Hydroponic gardening is a soilless method of cultivating plants by providing them with nutrient-rich water solutions. Hydroponic gardening is based on the fundamental concepts of supplying plants with vital nutrients, oxygen, water, and light to facilitate their growth. Below are a few key principles of hydroponic gardening:

- *Nutrient Solution:* Hydroponic systems utilize a nutrient solution that contains vital minerals necessary for plant development. The essential nutrients encompass nitrogen, phosphorus, potassium, calcium, and magnesium, as well as trace elements such as iron, manganese, zinc, and copper.

- *Water and Oxygen:* Hydroponically grown plants are immersed or cultivated in a medium that allows their roots to maintain continuous contact with the nutrition solution, ensuring sufficient levels of oxygen for root respiration. Hydroponic systems specifically engineer the root zone to receive sufficient levels of oxygen, which is vital for the process of root respiration. Methods for oxygenation encompass the utilization of air pumps, air stones, or oxygen diffusers inside the nutritional solution.

- **pH and EC Levels:** pH denotes the degree of acidity or alkalinity of the nutritional solution. Various plants have distinct pH preferences; nonetheless, a pH range of 5.5 to 6.5 is generally appropriate for the majority of hydroponic plants. Electrical conductivity (EC) quantifies the level of dissolved ions in the nutrition solution. It denotes the concentration of nutrients. Ensuring optimal pH and EC levels is essential for promoting healthy plant growth.

- *Lighting:* Plants necessitate light for photosynthesis, the biochemical process through which they transform light energy into chemical energy to support their growth. Hydroponic farming uses artificial lighting systems, such as LED, fluorescent, or HID (High-Intensity Discharge) lamps, to supply plants with the precise light spectrum required for their maximum growth.

- *Temperature and humidity:* are crucial factors for ensuring the well-being of plants in a hydroponic environment. The optimal temperature range is often from 65°F to 80°F (18°C to 27°C), and it is important to regulate humidity levels to minimize the formation of mold and fungi while facilitating transpiration.

- *Hygiene and Disease Control:* Maintaining adequate sanitation and practicing proper hygiene are crucial in hydroponic systems to prevent the transmission of diseases and infections. Regular maintenance of equipment, vigilant monitoring of nutrient solutions, and meticulous upkeep of a sterile growing environment effectively mitigate the potential for plant infections.

- *System Selection:* A range of hydroponic systems are accessible, encompassing deep water culture (DWC), nutrient film technology (NFT), aeroponics, ebb and flow (flood and drain), and drip systems. Each system has advantages and is appropriate for a variety of plant types and growth conditions.

Hydroponic gardeners can establish a conducive environment for plants to flourish and generate abundant crops without relying on soil by adhering to these fundamental principles.

CHAPTER 2

Understanding the components of hydroponics

Hydroponics is a soilless method of cultivating plants where a water solution containing vital nutrients is utilized to directly nourish the plant roots. The primary constituents of a hydroponic system generally encompass:

> **Growing medium:** While hydroponic systems do not depend on soil, they frequently employ a growing medium to uphold the plant roots and ensure stability. Commonly used growing

mediums include perlite, vermiculite, rockwool, coco coir, clay pellets, and sand.

➢ **Reservoir:** The reservoir is the receptacle that contains the nutrition solution. The reservoir can consist of various materials, including plastic, glass, or metal. The reservoir must be opaque to avoid the proliferation of algae and to preserve the integrity of the nutrient solution.

➢ **Nutrient Solution:** The nutrient solution directly supplies a blend of water and vital nutrients to the plant roots. The nutrition solution commonly comprises nitrogen, phosphorus, potassium, calcium, magnesium, and many trace elements that are required for plant development.

> **Water Pump:** The water pump is responsible for facilitating the circulation of the nutrient solution from the reservoir to the roots of the plants. It guarantees the continuous provision of nutrients and oxygen to the plants.

> **Aeration System:** The aeration system oxygenates the nutritional solution, hence supplying oxygen to the plant roots. Aeration can be achieved by using air stones, air pumps, or alternative techniques that facilitate oxygen exchange in the water.

> **pH and Electrical Conductivity (EC) Monitoring Equipment:** Hydroponic systems

necessitate careful monitoring of pH levels and electrical conductivity to guarantee that the nutrient solution remains within the appropriate range for plant growth. pH and EC monitoring equipment are employed for this purpose.

➢ **Grow lights:** are commonly employed in indoor hydroponic systems to supply the essential light spectrum required for plant photosynthesis. Farmers frequently select LED grow lights, fluorescent lights, and high-intensity discharge (HID) lights for indoor hydroponic systems.

➢ **Planting trays or containers:** The plants and the growing media are housed in planting trays or containers. The exact design of the hydroponic system and the sorts of plants being cultivated determine the shapes and sizes of the planting trays or containers.

➢ **Timers and controllers:** are automation components that aid in the management of hydroponic systems by precisely managing the timing and length of nutrient solution delivery, lighting cycles, and other crucial features.

➢ **Support Structures:** Certain hydroponic systems may necessitate the use of support structures, such as trellises or posts, to provide assistance in bearing the weight of the plants during their growth.

The components may differ based on the particular hydroponic system employed (e.g., deep water culture, nutrient film technology, aeroponics, etc.) and the specific requirements of the cultivated plants.

CHAPTER 3

Establishing a personal hydroponic system

Establishing your own hydroponic system can be an interesting and fulfilling endeavor. Hydroponics is a method of cultivating plants without soil where nutrient-rich water solutions are used to directly supply the plant roots with the necessary elements. A simple guide to help you get started is provided below:

1. Select a hydroponic system design:

Fundamental Systems:

- **Deep Water Culture (DWC):** is a hydroponic system where plants are held in a nutrient solution and their roots are fully submerged.

- ***The Nutrient Film Technique (NFT):*** involves the continuous flow of a nutrient solution over the plant roots in a thin film.

- ***The Ebb and Flow (Flood and Drain):*** This system involves periodically flooding the plant roots with a nutrient solution.

- ***Aeroponics:*** It is a cultivation method where plant roots are suspended in the air and regularly sprayed with a nutritional solution.

Factors to take into account:

- The availability of space

- Allocation of funds

29

- Quantity and categorization of desired plant species

- Desired level of automation

2. Collect essential supplies:

- A container or receptacle designed to contain the nutrient solution

- Growing mediums (for example, perlite, vermiculite, and coconut coir)

- Nutrient Solution

- pH testing kit and pH adjusters Grow trays or pots Pump (if necessary).

- Conduits and connectors

- Artificial lighting systems (if cultivating plants indoors) Timer (for automated systems).

- Seeds or young plants

3. Get Everything Set Up:

- *Container Setting Up:* Arrange your selected hydroponic system in accordance with its specific design. Make sure that the container or reservoir is both clean and lightproof in order to prevent the growth of algae.

- *Nutrient Solution:* Create your nutrient solution in accordance with the instructions provided by the manufacturer or a reliable recipe. Regularly monitor the pH levels and make necessary adjustments to maintain the ideal range (often between 5.5 and 6.5) for most plants.

- *Planting:* Sow your seeds or seedlings in the growing medium and place them in the system. Ensure that the roots are in direct contact with the nutrition solution.

- *Lighting:* Ensure sufficient illumination for your plants while cultivating them inside. Hydroponic systems widely use LED grow lights due to their high energy economy and spectrum adjustability.

- *Monitoring and Maintenance:* Conduct periodic assessments of pH and nutrient levels to maintain the utmost well-being of the plants. Observe carefully for any indications of pests or diseases. Modify the concentration and makeup of the nutrient solution as the plants develop.

4. Care and Maintenance:

- *Nutrient Solution Management:* Routinely monitor and modify the nutrient solution to uphold optimal pH and nutrient concentrations.

- *Lighting:* Ensure plants receive sufficient light exposure in accordance with their individual needs.

- *Pruning and Training:* Remove unnecessary foliage and trim plants to maximize space utilization and even dispersion of light.

- ***Pest and Disease Control:*** Observe plants for indications of pests or diseases and implement suitable measures to avert infection.

- ***Cleaning:*** Regularly clean the system and reservoir to prevent algae and bacterial growth.

5. Harvesting and Savoring Your Produce: Harvest your hydroponically grown plants when they are ready. Enjoy fresh, wholesome produce grown right at home!It is advisable to adopt a strategy of first starting with a small scale and progressively increasing the size as you acquire more expertise in the field of hydroponics.

CHAPTER 4

Picking the right veggies for hydroponic cultivation

Hydroponic cultivation is the practice of growing plants in a controlled environment where the roots are submerged in a solution that is rich in nutrients, instead of using soil. This technique enables meticulous regulation of crucial variables, including pH levels, nutritional concentrations, temperature, and humidity. Hydroponic systems enhance plant development and improve crop health by directly supplying all essential nutrients to the roots, resulting in faster growth rates and healthier crops compared to traditional soil-based approaches.

Important Considerations for Choosing Vegetables in Hydroponics

GROWTH NEEDS:

- *Light Intensity:* The majority of vegetables demand sufficient light in order to carry out photosynthesis. Hence, it is crucial to choose veggies that flourish in indoor settings with artificial illumination.

- *Temperature:* Vegetables suited for hydroponic

production must be able to withstand the temperature ranges generally maintained in indoor environments, which usually fall between 65°F and 80°F (18°C and 27°C).

- *Humidity:* Hydroponic systems have the ability to provide precise humidity levels that are favorable for plant growth. Nevertheless, selecting veggies that can easily adjust to different levels of humidity is beneficial.

REQUIRED NUTRIENTS:

- *Macronutrients:* Vegetables exhibit variability in their specific macronutrient needs, encompassing nitrogen, phosphorus, and potassium.

Comprehending the specific nutrient requirements of certain vegetables is essential for maintaining a well-balanced nutrient solution.

- *Micronutrients:* such as iron, calcium, and magnesium, are crucial for plant development. Choosing vegetables that exhibit a favorable response to the addition of micronutrients guarantees robust development and abundant harvests.

RATE OF GROWTH AND DURATION OF HARVEST:

- *Fast-growing cultivars:* Certain vegetables, including lettuce, spinach, and herbs, possess rapid growth cycles and are well-suited for hydroponic systems owing to their capacity to yield multiple harvests within a brief timeframe.

- *Long-Growing Varieties:* Although vegetables such as tomatoes and cucumbers require more time to mature, their ability to produce a large amount of crops and their ability to thrive in hydroponic conditions make them acceptable options for lengthy culture periods.

OPTIMIZATION OF SPACE AND UTILIZATION OF RESOURCES:

- *Compact Growth Habit:* In hydroponic systems with limited space, it is best to choose vegetables with compact growth habits and low space requirements.

- **Water Efficiency:** Selecting plants that demonstrate efficient water utilization can aid in conserving resources and maximizing the sustainability of hydroponic systems.

Ideal Vegetables for Hydroponic Cultivation

LEAFY GREENS:

- *Lettuce:* Butterhead, romaine, and leaf lettuce thrive in hydroponic systems with rapid growth rates and multiple harvests.

- *Spinach:* Because of its dense growth habit and high nutrient density, spinach grows well in hydroponic systems.

- *Kale:* Kale varieties thrive in hydroponic environments, producing nutrient-dense leaves that can be continuously harvested.

HERBS:

- *Basil:* One of the most common herbs grown in hydroponic systems, basil provides aromatic foliage and grows quickly when given the right conditions.

- *Cilantro:* It is suitable for hydroponic systems as it produces fresh leaves that can be used in cooking throughout its entire growth cycle.

- *Parsley:* Parsley flourishes in hydroponic systems and provides a consistent abundance of savory leaves for culinary purposes.

VINE CROPS:

- *Tomatoes:* Compact tomato cultivars, such as cherry and grape tomatoes, are ideal choices for hydroponic farming due to their ability to produce abundant crops and deliver superb taste.

- *Cucumbers:* They thrive in hydroponic systems, yielding plentiful fruit on little vines that are ideal for vertical cultivation.

PEPPERS:

- ***Bell peppers:*** They thrive in hydroponic conditions, providing vivid hues and crunchy textures for culinary purposes.

- ***Chili Peppers:*** Different types of chili peppers, such as jalapenos and habaneros, grow quite well in hydroponic systems, producing hot flavors and abundant harvests.

Choosing the appropriate vegetables is crucial for the success of hydroponic cultivation endeavors. Growers can optimize the efficiency and long-term viability of their hydroponic systems by taking into account elements such as growth requirements, nutrient needs, and practical issues. Selecting vegetables that are well-suited for hydroponic conditions is crucial for achieving optimal development, high yields, and superior quality food for culinary and commercial purposes. This applies to leafy greens, herbs, vine crops, and peppers. The continual improvement of hydroponic technology has great potential for the development of novel vegetable production techniques, which in turn can contribute to a more environmentally friendly and sustainable future in agriculture.

CHAPTER 5

Preparing a nutrient solution for your hydroponic garden

The process of preparing a nutrient solution for hydroponics entails precisely measuring and blending vital nutrients into water to form a well-balanced solution that supplies plants with all the necessary elements for robust growth. Simply follow these simple steps to make a basic nutrient solution for hydroponics:

KNOW THE NUTRIENT REQUIREMENTS: Plants need many vital elements for healthy development, such as nitrogen (N), phosphorus (P), potassium (K), calcium

(Ca), magnesium (Mg), sulfur (S), iron (Fe), manganese (Mn), zinc (Zn), copper (Cu), molybdenum (Mo), and boron (B).

CHOOSE NUTRIENT SOURCES: You have the option to obtain certain nutritional salts or compounds in order to construct a personalized nutrient solution. Frequent sources include:

- Calcium nitrate (Ca(NO3)2)

- Potassium nitrate (KNO3)

- Magnesium sulfate (MgSO4)

- Monopotassium phosphate (KH2PO4)

- Iron chelates (such as EDTA or DTPA) Trace element mixes (containing micronutrients like zinc, manganese, copper, etc.)

COLLECT THE NECESSARY MATERIALS:

- Hydroponic nutrient solution blend (available for purchase or made at home)

- Pure water (preferably distilled or obtained through reverse osmosis)

- Scales or cups for measuring

- pH meter or pH testing kit pH-up and pH-down solutions for pH adjustments.

- Electrical conductivity (EC)/parts per million (PPM) meter (optional but highly recommended)

IDENTIFY NUTRIENT FORMULAS: Plants of

different species have diverse nutrient needs throughout different phases of growth. Conduct research or seek advice from professionals to ascertain the suitable nutrient formulas for your plants during different growth stages, such as vegetative, blooming, and fruiting.

CALCULATE NUTRIENT SOLUTION CONCENTRATION: Determine the preferred concentration of your nutrient solution by considering the plant's growth stage, plant variety, and surrounding environmental factors. To obtain guidance on the appropriate usage of your nutrient solution mix, you may consult the instructions that come with it or seek guidance from resources tailored to your plants' unique needs.

PREPARING THE NUTRIENT SOLUTION:

- Begin by acquiring a clean container or reservoir that possesses the capacity to accommodate the appropriate volume of nutritional solution.

- Fill the container with water, preferably distilled or reverse osmosis water, to minimize contaminants and guarantee uniformity.

- Sequentially introduce each nutrient salt or compound into the water, adhering to the predetermined concentrations for each nutrient. Ensure uniform dispersion of each nutrient by thoroughly stirring the fluid after adding it.

- Continuously measure the acidity level of the solution while introducing nutrients. Regulate the pH level by employing pH up (often potassium hydroxide or potassium carbonate) or pH down (usually phosphoric acid) solutions as necessary to achieve the ideal pH range for your plants (generally between 5.5 and 6.5 for most hydroponic crops).

- Utilize an EC/PPM meter to quantify the electrical conductivity, or parts per million, of dissolved salts in the solution. Modify the nutrient concentrations as needed in order to attain the required EC/PPM levels.

➢ Allow the nutrient solution to undergo a brief period of settling in order to establish stability prior to utilizing it for irrigation in your hydroponic system.

KEEP RECORDS: Maintain documentation of nutrient solution formulas, pH measurements, EC/PPM measurements, and any subsequent modifications. This aids in the identification and resolution of problems, as well as the preservation of uniformity in the distribution of nutrients.

LABEL AND STORE: Attach a label to the nutrient solution container indicating the date and nutrient concentrations in order to maintain a record of its composition. Preserve any leftover solution by storing it in a cold and dark location to avoid deterioration.

EVALUATE AND MODIFY: Regularly assess the nutrient solution and observe the condition of the plants to verify that the nutrient levels are suitable and that the plants are obtaining the essential ingredients required for robust development. Modify the nutrient solution as necessary, taking into account the plant's reaction and stage of growth.

To ensure optimal plant growth and abundant yields in your hydroponic system, it is crucial to carefully follow these instructions and consistently monitor your plants and nutrient solution. This will enable you to proficiently

formulate your own nutrient solution.

CHAPTER 6

Tips on how to cultivate plants

hydroponically

Hydroponic gardening is a method of growing plants without soil, using nutrient-rich water solutions instead. Listed below are some step-by-step instructions that can assist you in growing your plants using hydroponics.

Step 1: Select and begin your plants and seeds.

Prior to commencing a hydroponics farm, it is imperative to make a deliberate choice regarding the specific variety of plants you intend to cultivate. Your decision will dictate the type of hydroponics system you can employ.

Additionally, it assists in determining the appropriate lighting type and the required spatial coverage. Certain hydroponic setups or systems outperform others when it comes to specific types of plants. If you are a beginner cultivator, it is suggested that you cultivate leafy greens. Rapid rooter cubes are suitable for seed germination. It is my belief that these plugs will facilitate your process by improving the germination rates. Also, these plugs will enhance the development of a healthy root system.

Step 2: Select the hydroponic system of your choice.

To initiate your hydroponics farm, the next step involves choosing the hydroponics system. Various hydroponic systems are available for selection. Whatever option you select, be sure to invest some time in becoming knowledgeable about each technique.

Several factors influence the selection of the optimal hydroponics system:

➢ Plant species

➢ The size of the space

➢ The allocation of financial resources

➢ Your desired level of growth

It is advised that you begin on a small scale, as there is the possibility of expanding in the future. Exposure to an excessive number of plants simultaneously can be overwhelming during the learning process.

Step 3: Choose a light source.Full-spectrum LED grow lamp.

Optimal illumination is a crucial determinant for the success of your hydroponics garden. If you opt not to utilize sunlight as your primary light source, it is necessary to acquire a specialized grow lamp for your hydroponics system. It is recommended that you get a full-spectrum, high-efficiency LED white grow light. Several considerations arise while choosing the optimal lighting for your hydroponics system. The reason for this is the existence of various types of grow lights, each with its own set of pros and cons. Conduct thorough research to identify the optimal lighting configuration that is well suited for your garden.

When selecting a grow light, there are several factors that you must take into account:

> The area covered

> Spectrum of light

> The intensity of light

Avoid opting for the most economical lighting option. Commence farming by utilizing a full-spectrum LED grow light from the very beginning. They are inexpensive, capable of producing fruits and leafy greens, and are also relatively efficient.

Step 4: Select a growing medium.

When establishing your hydroponics garden, ascertain the specific type of growing medium you choose. Select the optimal system and growing medium based on the specific plant species you intend to cultivate. Various types of growing substrates provide multiple benefits. However, while selecting a growth medium, it is important to take into account pH stability, aeration levels, retained water, and expense.

Below are a few growth media you may want to explore:

➢ **Rockwool:** Basalt rock is melted and transformed into interconnected fibers through a spinning process to create Rockwool, a widely used growing medium. Rockwool exhibits a notable water-holding ability, ensuring that even in the event of a water pump failure, your plants would continue to receive water. Their high air-holding capacity prevents excessive water from reaching your plants.

> **Perlite:** Perlite is considered one of the most superior planting mediums for hydroponics. We attribute this to its inert characteristics, porosity, and lightweight nature. The process involves subjecting glass flakes (silica) to heat, causing them to expand in a manner similar to popcorn. This medium possesses exceptional water-absorption characteristics.

➢ **Coco coir:** Coconut fiber is commonly referred to as coco coir. The powdered coconut husks are a completely organic medium with significant water and oxygen retention capability. As a result, if there is any system malfunction, your plants will have a continuous chance of survival.

Step 5: Acquire dietary supplements and hydroponic nutrients.

Various types of nutrients are delivered in systems consisting of one, two, or three parts. For inexperienced cultivators, it is recommended that you use a Part One nutrition solution specifically designed to support the processes of flowering, blooming, and vegetative growth.

Various internet retailers offer a wide range of equipment, additives, and fertilizers specifically designed for hydroponic farming. Farmers can include various supplements to enhance the growth rate, flavor, and overall size of plants.

Step 6: Get a pH Up/Down and pH Meter.

A pH-up/down solution with a pH meter. Plants in a hydroponics system can assimilate nutrients within a specific pH threshold. Get a pH meter in order to frequently measure the pH of nutritional solutions. You have the option to acquire an electronic meter, test strips, and a liquid kit. It is necessary to purchase both a pH-down and a pH-up solution for adjusting the pH level of your nutrient solution. It is crucial to promptly address any sudden increases or decreases in nutrient levels to Effective planning is crucial for achieving a thriving hydroponic garden. Investing effort in thorough initial planning can yield significant long-term savings in both finances and time. Choose the most superior grow lights, obtain the optimal nutrient blend, select the finest hydroponics system, and go with your cultivation endeavors. I sincerely hope that these comprehensive instructions save you time and money while facilitating the success of your hydroponic garden. prevent nutrient lockout.

Transplanting seedlings into hydroponic

system

The process of transferring seedlings into a hydroponic system necessitates care and monitoring to guarantee the successful acclimatization of the juvenile plants to their unfamiliar surroundings. Below is a comprehensive guide to transferring seedlings into a hydroponic system:

SET UP YOUR HYDROPONIC SYSTEM:

- Make sure your hydroponic system is established and functioning correctly. Verify the adequacy of the pH level, nutrient solution, and water level to determine their suitability for the plants you are transferring.

GET THE SEEDLINGS READY:

- Choose healthy seedlings that have mature leaves and strong stems. Seedlings must be devoid of illnesses, pests, and other indications of stress. To transplant seedlings from the soil, carefully remove them from the soil, ensuring the roots remain unharmed.

- To eliminate dirt particles, you may either gently shake off any excess soil or rinse the roots with water.

TRIM THE ROOTS IF NEEDED:

- If you are utilizing a hydroponic system, it may be necessary to trim the roots of the seedlings in order to promote healthy development in the hydroponic environment. Trim any extensive or injured roots using sterile, sharp scissors.

PREPARE THE HYDROPONIC MEDIUM:

- Prior to use, it is essential to soak rockwool cubes or hydroponic clay pellets in water or fertilizer solution as directed by the manufacturer.

TRANSPLANT THE SEEDLINGS:

- Carefully position each seedling in its assigned location within the hydroponic system. Handle the roots carefully and refrain from bending or causing harm to them when carrying out the transplanting procedure.

- Ensure that the roots are completely immersed in the nutrient solution or in direct contact with the hydroponic media to guarantee optimal absorption of nutrients.

- Adjust the distance between seedlings to ensure sufficient air movement and space for expansion.

OBSERVE AND ADJUST:

- Following transplantation, diligently observe the seedlings for any indications of distress or insufficiency in nutrients. Monitor the water level, pH level, and nutrient content in the hydroponic system.

- Make any required modifications to ensure the

seedlings are in the most favorable growing environment.

PROVIDE SUPPORT (IF NECESSARY):

- Certain seedlings may necessitate support as they acclimate to their unfamiliar surroundings. Utilize stakes or trellises to provide support for tall or elongated plants, preventing their inclination or fracturing.

MAINTAIN CARE:

- Maintain optimal light, temperature, humidity, and nutrient conditions as the seedlings develop in the hydroponic system.

- Frequently inspect the roots and system for

indications of issues such as root decay, algae proliferation, or nutrient disparities, and promptly address them.

By adhering to these instructions and ensuring adequate maintenance, your seedlings should successfully acclimate to their new hydroponic setting and flourish as they progress into fully developed plants.

CHAPTER 7

Maintaining your hydroponic garden

Maintaining a hydroponic garden requires several critical actions to maintain healthy plant growth and a good harvest. Here's a general guide to help you maintain your hydroponic garden:

1. Monitor Nutrient Solution: Consistently assess the pH and nutrient concentrations of your hydroponic solution by employing a pH meter and a TDS (Total Dissolved Solids) meter. Adjust the nutrient solution as needed to maintain the optimal pH and nutrient concentration for your plants.

2. Monitor Water Levels: Check that your hydroponic system maintains a sufficient amount of water consistently. Monitor water levels in the reservoir and top up as necessary to prevent the roots from drying up.

3. Monitor Plant Health: Regularly monitor your plants for symptoms of pests, illnesses, nutritional deficiencies, or other difficulties. To mitigate the potential impact on the entire crop, swiftly address any issues that arise to preve

4. Prune and Trim: Trim back excess foliage and prune plants as needed to ensure healthy growth and prevent overcrowding. This enhances air circulation and facilitates the penetration of light, hence diminishing the likelihood of sickness and fostering more efficient nutrient absorption.

5. Ensure Optimal Lighting: If you are utilizing artificial lighting in your hydroponic system, make sure that the lights are in perfect working condition and positioned at the appropriate distance from the plants. Replace bulbs as necessary and maintain cleanliness of light fixtures to optimize light emission.

6. Maintain Hygiene: Routinely cleanse and disinfect your hydroponic system to avert the accumulation of algae, bacteria, and other harmful microorganisms. This includes cleaning grow trays, reservoirs, pumps, tubing, and any other components of your system.nt their propagation.

7. Monitor Environment Condition: Ensure that you maintain the temperature and humidity levels in your hydroponic grow space at the ideal levels to provide an optimal growing environment for your plants. Utilize fans, heaters, humidifiers, and dehumidifiers as necessary to control and adjust the surrounding environment.

8. Ensure Proper Air Circulation: Sufficient air circulation is crucial for promoting optimal plant growth in a hydroponic system. Employ fans or ventilation devices to guarantee enough air circulation around your plants.

9. Crop Harvesting and Rotation: Consistently gather

fully grown plants to stimulate regrowth and sustain an uninterrupted provision of recently harvested produce. Implement crop rotation throughout successive cultivation periods to mitigate nutrient depletion and reduce the likelihood of soil-borne illnesses.

10. Maintain Records: Maintain careful records of your hydroponic garden, encompassing nutrient regimens, pH levels, plant growth rates, and any complications encountered. This data will enable you to discern patterns, diagnose issues, and enhance your cultivation methods progressively.

By adhering to these guidelines and maintaining an active approach to the upkeep of your hydroponic garden, you can guarantee robust plant development and a fruitful yield.

CHAPTER 8

Troubleshooting common problems

The popularity of hydroponic gardening has surged because of its resource efficiency, capacity to cultivate plants in small spaces, and superior yields in comparison to conventional soil-based techniques. Nevertheless, similar to any other form of cultivation, hydroponics is not devoid of its own set of obstacles. Cultivators frequently face diverse challenges that can hinder plant development and output. It is essential to comprehend and resolve these prevalent hydroponic problems in order to sustain vigorous plants and attain maximum yields.This guide will look at common problems that arise in hydroponic systems and provide practical strategies to efficiently resolve them.

➢ Nutrient Imbalance

An essential part of hydroponics is ensuring that plants receive proper nutrients in the appropriate ratios. Nutrient imbalances may occur as a result of factors such as inappropriate nutrient blending, variations in pH levels, or insufficient nutrient absorption by plants. Indications of nutritional imbalance encompass the occurrence of yellowing leaves, hindered growth, and leaf discoloration.

Solution:

- Consistently check nutrient levels by utilizing an electrical conductivity (EC) meter and a pH meter.

- Adhere to the manufacturer's instructions for the proportions of nutrients to be mixed, and make any required adjustments depending on the stage of plant growth.

- Regularly flush the system to avoid the accumulation of salt, which may interfere with the absorption of nutrients.

- Maintain the pH levels within the ideal range of 5.5 to 6.5 for individual plant species.

- If shortages are identified, it is advisable to utilize a well-balanced nutrient solution or incorporate a micronutrient supplement.

➢ **Root Rot**

Pathogenic fungi and bacteria primarily cause root rot, a prevalent problem in hydroponic systems. It is particularly common in settings characterized by high humidity and inadequate oxygenation. Indications comprise mucilaginous, dark roots, unpleasant smells, and withering vegetation. If not handled, root rot can have a substantial influence on the absorption of nutrients and the overall health of plants.

Solution:

- Ensure enough oxygenation and aeration in the hydroponic system by employing air stones, air pumps, or oxygen injectors.

- Ensure the water temperature remains within the ideal range of 65°F to 75°F or 18°C to 24°C to prevent the proliferation of pathogens.

- Employ sterilized growing media and utilize clean, sanitized equipment to minimize the entry of germs.

- Establish a proactive treatment protocol by employing advantageous bacteria and fungi to overcome detrimental diseases.

- It is advisable to include hydrogen peroxide or other natural antifungal medicines to manage the proliferation of microorganisms in the root area.

> **Temperature Fluctuations**

Temperature variations can exert a substantial influence on the development of plants, the absorption of nutrients, and the overall well-being of hydroponic systems. Severe temperatures can cause plants to experience stress, decrease the solubility of oxygen, and facilitate the growth of diseases and algae.

Solution:

- Ensure that the temperature in the growing environment is kept at an ideal level to promote the healthy growth and metabolic processes of plants.

- Employ temperature regulation devices such as heaters, chillers, or evaporative coolers to control water and air temperatures.

- Implement temperature sensors and monitoring systems to accurately measure temperature variations and promptly detect any problems.

- Ensure sufficient ventilation and air movement to disperse heat and maintain consistent temperatures across the entire cultivation area.

- It is advisable to utilize thermal insulation materials in order to maintain stable temperature levels and minimize energy usage.

> **pH Fluctuations**

Fluctuations in pH can have a substantial impact on the availability of nutrients and the ability of plants to absorb them. Various factors, including the concentration of nutrients, the quality of water, and the activity of microorganisms, can affect the pH levels in hydroponic systems. Deviation from the ideal range can cause nutrient lockout, a condition in which vital nutrients become inaccessible to plants, leading to nutrient shortages.

Solution:

- Consistently check the pH levels by utilizing a dependable pH meter, and make necessary adjustments by employing pH up or pH down solutions.

- Maintain a consistent nutrition solution and employ pH stabilizers, if needed, to regulate pH changes in the buffer.

- It is advisable to include pH buffering agents or additives in order to stabilize the pH values in the reservoir.

- Increase the frequency of monitoring and adjusting pH levels during periods of accelerated plant development or fluctuations in the environment.

> **Algae growth**

Hydroponic systems exposed to light, particularly in nutrient reservoirs and growing media, commonly face the issue of algae growth. Algae engage in competition with plants for nutrients and oxygen, resulting in diminished nutritional availability and heightened susceptibility to root rot. Excessive proliferation of algae can obstruct irrigation pipes and compromise the functioning of the system.Solution:

- Minimize the exposure of nutrient reservoirs and growing media to light by utilizing containers that are opaque or made of light-blocking materials.

- Ensure adequate water movement and aeration to reduce the presence of stagnant water and oxygenate the nutritional solution.

- To effectively manage the growth of algae without causing harm to plants, utilize algae inhibitors or algaecides that are formulated with natural components.

- Initiate frequent cleaning and maintenance procedures to eliminate the accumulation of algae on system components.

- It is advisable to include UV sterilizers or ozone generators to prevent the formation of algae and ensure the maintenance of water quality.

Achieving success in hydroponic gardening necessitates extreme attention to detail and proactive control of any system-related concerns. To ensure ideal growing conditions and generate abundant harvests, gardeners can overcome frequent obstacles in hydroponic agriculture by comprehending them and employing suitable troubleshooting approaches. Hydroponic gardeners can optimize their indoor gardening efforts by swiftly and successfully managing nutrient imbalances, pH changes, temperature swings, root rot, and algae growth. By demonstrating dedication, patience, and steadfast adherence to optimal procedures, hydroponic enthusiasts can experience the gratification of flourishing plants and long-lasting sustainable production.

CHAPTER 9

Harvesting and enjoying your produce

Engaging in the process of gathering and savoring the crops in a hydroponic system may be a pleasant experience.

Knowing when to harvest

Accurately determining the appropriate time to gather crops in a hydroponic system is essential to guaranteeing the highest quality in terms of taste, consistency, and nutritional content. Below are some universal principles to assist you in determining the optimal time for harvesting different crops in your hydroponic system:

- *Study Plant Tags or Seed Packets:* Begin by acquainting yourself with the precise type of crop you intend to cultivate. Plant tags or seed packs commonly include details regarding the average duration until the plant reaches maturity and the typical attributes it possesses when it is ready to be harvested.

- *Monitor Growth and Development:* Monitor the growth and development of your plants on a regular basis to observe their progress. Monitor and record characteristics such as leaf dimensions, pigmentation, and the overall visual presentation of the plant. To monitor the progress of fruiting crops such as tomatoes and peppers, carefully examine the size, color, and hardness of the fruits during their growth.

- *Check Ideal Harvest Periods:* Numerous hydroponic crops possess suggested harvest periods determined by the number of days required for maturity or certain developmental phases. To ascertain the ideal time to harvest each product, one should check gardening materials, hydroponic guides, or seek advice from experienced gardeners.

- *Conduct Harvest Tests:* To determine if particular crops, including leafy greens and herbs, are ready for harvesting, you can carry out harvest tests. Harvest a small sample of leaves or stems and evaluate their taste, texture, and scent. If the remaining crop meets your specified quality standards, it is probably the appropriate time to gather the rest of it.

- *Observe Visual Indicators:* Look for visual cues that indicate the plant is ready for harvest. Lettuce leaves should fully grow and exhibit a vivid hue, while herbs should possess a strong aroma and appear luxuriant. Fruits should attain their distinctive color and size, and they should possess a strong texture when touched.

- *Take into account environmental factors:* Factors such as temperature, humidity, and light intensity can have an impact on the growth and development of plants. Revise your harvesting timetable appropriately, considering any variations in environmental circumstances that can impact the ripening of crops.

- *Optimal Harvest Timing:* It is advisable to harvest crops during the early hours of the day when they are adequately hydrated and have not been subjected to excessive heat or sunlight. This aids in maintaining the integrity and freshness of the harvested crops.

- *Implement Crop Rotation:* In order to ensure a consistent harvest, it is advisable to strategically schedule plants and routinely rotate crops inside your hydroponic system. This guarantees a consistent provision of newly harvested crops and avoids excessive population density that might hinder the progress and maturation of plants.

To optimize the flavor and quality of your hydroponic vegetables, it is crucial to employ these tactics and remain attentive to their specific requirements. This will enable you to harvest them in their best state, ensuring maximum enjoyment and nutritional advantages.

Harvesting techniques for different vegetables

The harvesting methods for various vegetables in a hydroponic system may differ depending on the specific plant species and its unique growing characteristics. Below are several prevalent harvesting methods for different vegetables cultivated in hydroponic systems:

LEAFY GREENS SUCH AS LETTUCE, SPINACH, KALE, SWISS CHARD, AND OTHERS:

- Harvest the outer leaves individually as required, while leaving the interior leaves to keep growing.

- Utilize sterile scissors or pruning shears to trim leaves in close proximity to the plant's base, ensuring to allow a few inches of growth to stimulate subsequent development.

- Frequently collect fully developed foliage to stimulate fresh sprouting and avoid excessive density inside the hydroponic setup.

HERBS SUCH AS BASIL, MINT, CILANTRO, PARSLEY, AND OTHERS:

- Pluck individual leaves or stems as required, commencing with the uppermost part of the plant.

- Promptly remove flower buds to promote leaf growth and prevent the plant from undergoing bolting (the process of producing seeds).

- For the best flavor, it is advisable to gather herbs in the morning, as this is when their essential oils are most concentrated.

TOMATOES:

- Collect tomatoes when they reach their mature hue (red, yellow, orange, etc.) and firmness.

- Utilize precise scissors or pruning shears to sever the stem directly above the fruit cluster.

- Exercise caution when handling tomatoes to prevent any bruising or harm to the fragile fruits.

PEPPERS (SUCH AS BELL PEPPERS, CHILI PEPPERS, AND OTHERS):

- Harvest peppers once they have attained the proper dimensions and hue.

- Utilize a pair of sharp scissors or pruning shears to precisely sever the stem directly above the fruit.

- To prevent harm to the stems and adjacent fruits, refrain from yanking or rotating the peppers off the plant.

CUCUMBERS:

- Harvest cucumbers when they are firm, have a vibrant color, and have reached the ideal length.

- Utilize precise scissors or pruning shears to sever the stem directly above the fruit, ensuring that a small section remains connected to the cucumber.

- To ensure a continual yield of cucumbers and avoid excessive ripening, it is advisable to harvest them on a regular basis.

STRAWBERRIES:

- Harvest strawberries when they are completely mature and have acquired their distinctive color and fragrance.

- Carefully hold the stem located above the fruit and utilize sharp scissors or pruning shears to sever it.

- To avoid excessive ripening and promote the growth of fresh fruits, it is advisable to regularly gather strawberries throughout the height of their growing season.

MICROGREENS:

- Microgreens typically emerge their initial true

leaves and can be harvested within 1-3 weeks after seed sowing.

- Utilize sterile scissors to precisely trim microgreens at the point immediately above the surface of the soil.

- Harvest microgreens in small quantities as required for immediate consumption or preservation.

By utilizing these harvesting methods, you can guarantee the utmost quality, taste, and efficiency of your hydroponically cultivated veggies during the whole growth period.

Storing and preserving harvested

produce

Properly managing harvested produce from a hydroponic system is essential for retaining its freshness and quality. Below are guidelines for storing and conserving harvested hydroponic produce:

- **Optimal Timing for Harvest:** Gather your hydroponic crops when they have reached their maximum level of ripeness. This guarantees optimal taste and nutritional value.

- **Handle with care:** Treat the gathered crops delicately to prevent any bruising or harm to the fragile fruits and vegetables.

- **Purify and Disinfect:** Thoroughly cleanse the harvested crops using clean water to eliminate any soil, fragments, or contaminants. Disinfect the containers or bins utilized for storage to inhibit the proliferation of bacteria and fungi.

- **Eliminate Excessive Moisture:** An abundance of moisture can facilitate the proliferation of mold and germs. Prior to storage, it is advisable to either air-dry the collected produce or carefully blot it dry using a clean towel.

- **Optimal Temperature and Humidity:** Various types of produce possess distinct temperature and humidity needs for storage. Store hydroponically grown crops in a cool, dry location with sufficient ventilation. Nevertheless, certain crops, like leafy greens, have a preference for slightly elevated humidity levels.

- **Utilize Appropriate Containers:** Preserve the gathered crops in breathable receptacles such as perforated plastic bags, mesh bags, or containers equipped with ventilation apertures. Do not store produce in sealed bags since it can result in decay.

- **Monitor and Rotate:** Periodically inspect the stored produce for indications of spoilage or deterioration. Rotate the produce to promote uniform air circulation and prevent anything from being compressed or damaged.

- **Prevent Ethylene Exposure:** Certain fruits and vegetables emit ethylene gas, which can hasten the ripening and deterioration of other products. To avoid premature deterioration, it is important to segregate ethylene-producing goods from ethylene-sensitive vegetables.

- **Refrigeration:** Refrigeration is beneficial for certain hydroponically cultivated crops, like leafy greens and herbs, as it helps prolong their freshness and durability. To preserve their freshness, place them in the crisper drawer of the refrigerator.

- **Implement Preservation Techniques**: In the event that you possess an excess of gathered produce, contemplate preserving it through means such as freezing, dehydrating, or canning. Implementing these strategies can effectively extend the longevity of your hydroponic yield.

By adhering to these guidelines, you may efficiently store and maintain your harvested hydroponic produce, guaranteeing its freshness and quality for an extended period.

Acknowledgement

The Glory of this book's success goes to God Almighty and my ever-loving Family, Fans, Readers & well-wishers, Customers, and Friends for their endless support and encouragement.

ABOUT THE AUTHOR

Rita van Klaveren is an enthusiastic advocate in the field of agriculture and sustainable farming methods. Rita's extensive practical expertise in land management and crop cultivation has provided her with a profound comprehension of the intricate and demanding nature of contemporary agriculture. She embarked on her agricultural path at her family's farm, where she acquired an understanding of the importance of diligence, perseverance, and responsible management of the land. Rita's interest in hydroponics started during her undergraduate studies in environmental science, where she extensively explored the fundamentals of hydroponic systems and their capacity to transform conventional farming methods. Subsequently, she has committed her professional life to investigating the complex interplay of nutrient solutions, root-zone habitats, and regulated growing conditions. Rita's work, as a farmer and prolific writer, extends beyond the boundaries of academia, reaching farmers, entrepreneurs, and individuals around the globe. Through her perceptive publications and captivating lectures, she elucidates the intricacies of hydroponic farming, enabling people to utilize its revolutionary capacity. Rita's experience in hydroponics has enabled her to develop successful and sustainable systems for growing high-yield crops throughout the year, ranging from small urban gardens to huge commercial enterprises. Her methods prioritize environmental conservation. Her comprehensive strategy prioritizes sustainability, optimizing resource utilization, and promoting a wide range of crops, thus establishing a more resilient and food-secure future. Amidst a time characterized by increasing worries about the availability of food and the preservation of the environment, Rita emerges as a symbol of creativity and optimism, advocating for a concept of farming that goes beyond conventional limitations. With her steadfast commitment and limitless creativity, she persistently challenges the limits of hydroponic farming, leading the path towards a more environmentally friendly and bountiful future.